SYSTEMS

of

EQUATIONS

Substitution, Simultaneous, Cramer's Rule

Algebra Practice Workbook with Answers

Improve Your Math Fluency Series

$$3x - 4y = 8$$

$$2x + 5y = 13$$

Chris McMullen, Ph.D.

Systems of Equations
Substitution, Simultaneous, Cramer's Rule
Algebra Practice Workbook with Answers

Improve Your Math Fluency Series

Zishka Publishing

Professional & Technical > Science > Mathematics > Algebra
Education > Specific Skills > Mathematics > Algebra

ISBN-10: 1-941691-04-8
EAN-13: 978-1-941691-04-5

Contents

Making the Most of this Workbook iv

Chapter 1: Substitution Method 5

Chapter 2: Substitution with 3 Unknowns 25

Chapter 3: Simultaneous Equations 53

Chapter 4: 2×2 Determinants 77

Chapter 5: Cramer's Rule 91

Chapter 6: 3×3 Determinants 111

Chapter 7: Cramer's Rule with 3 Unknowns 127

Appendix: Solving Systems with a TI-83 Calculator 161

Answer Key 169

Making the Most of this Workbook

Mathematics is a language. You can't hold a decent conversation in any language if you have a limited vocabulary or if you are not fluent. In order to become successful in mathematics, you need to practice until you have mastered the fundamentals and developed fluency in the subject. This workbook will help you improve the fluency with which you solve systems of equations for two or more unknowns.

This workbook covers three different common techniques for solving two (or more) equations in two (or more) unknowns: the method of substitution, simultaneous equations, and Cramer's rule.

Each chapter begins with a concise explanation of the strategy in words along with an example. Use the example as a guide until you become fluent in the technique.

After you complete a page, check your answers with the answer key in the back of the book. Practice makes permanent, but not necessarily perfect: If you practice making mistakes, you will learn your mistakes. Check your answers and learn from your mistakes such that you practice solving the problems correctly. This way your practice will make perfect.

Chapter 1

Substitution Method

Concepts

Two equations with two unknowns can be solved via the method of substitution.

The main idea is to isolate one unknown in one equation and then plug this expression into the unused equation.

Consider the system of equations below.

$$3\,x - 2\,y = 8$$
$$5\,x + 4\,y = 6$$

This system has two equations. Each equation has two unknowns, x and y.

One way to solve the system is to isolate x in the first equation and plug the resulting expression in for x in the second equation.

The result is one equation with just one unknown (y), which can be found by combining like terms.

Strategy in Words

Step 1: First, isolate one unknown (x or y) in one equation.

> **Isolate** means to have that variable on one side of the equation all by itself.

Step 2: Then substitute the expression for that unknown into the unused equation.

> **Tip**: Be careful not to plug the expression back into the equation that you already used.

Step 3: You should now have one equation with just one unknown. Isolate that unknown to solve for it.

Step 4: Once you solve for one unknown, plug it back into any equation to find the other unknown.

Guided Example

Solve the following system for x and y.

$$3x - 2y = 8$$
$$5x + 4y = 6$$

Note that $8 + 2y$ is the same as $2y + 8$

Step 1: Isolate x in the first equation.

Do this by adding $2y$ to both sides, and then dividing both sides by 3.

$$3x = 2y + 8$$
$$x = \frac{2y + 8}{3}$$

Step 2: Plug this expression in for x in the bottom equation.

Place parentheses around the expression when you plug it into the bottom equation.

This equation came from above

Plug x in here

$$5\left(\frac{2y + 8}{3}\right) + 4y = 6$$

Step 3: Solve for y in this equation.

Distribute the 5.

Note that $5\left(\frac{2y+8}{3}\right) = \frac{10y}{3} + \frac{40}{3}$

$$\frac{10y}{3} + \frac{40}{3} + 4y = 6$$

Group like terms together. Put all the y-terms on one side of the equation and the constant terms on the other side.

$$\frac{10\,y}{3} + 4\,y = 6 - \frac{40}{3}$$

To combine like terms, you must add fractions. This requires finding a common denominator (i.e. $4\,y = \frac{12\,y}{3}$ and $6 = \frac{18}{3}$).

$$\frac{10\,y}{3} + \frac{12\,y}{3} = \frac{18}{3} - \frac{40}{3}$$
$$\frac{22\,y}{3} = -\frac{22}{3}$$

In order to isolate y, multiply both sides by 3 and divide both sides by 22.

$$y = -1$$

Step 4: Plug this value for y into any equation that has both x and y. The simplest choice is the equation where x is already isolated.

$$x = \frac{2\,y}{3} + \frac{8}{3}$$
$$x = \frac{2\,(-1)}{3} + \frac{8}{3} = -\frac{2}{3} + \frac{8}{3} = \frac{6}{3} = 2$$

Note that $\frac{2\,y}{3} + \frac{8}{3}$ is the same as $\frac{2\,y + 8}{3}$

The final answers are:

$$x = 2 \text{ and } y = -1$$

Check Your Answers

You can check your answers by plugging x and y into the original equations.

$$3x - 2y = 8$$
$$5x + 4y = 6$$

First plug $x = 2$ and $y = -1$ into the top equation.

$$3(2) - 2(-1) = 8$$
$$6 + 2 = 8$$
$$8 = 8$$

Since $8 = 8$, we know that x and y satisfy the first equation.

Now plug $x = 2$ and $y = -1$ into the bottom equation.

$$5(2) + 4(-1) = 6$$
$$10 - 4 = 6$$
$$6 = 6$$

Since $6 = 6$, x and y also satisfy the bottom equation.

Solve for x and y in each system using substitution.

Problem 1

$$8\,x + 9\,y = -\,94$$

$$8\,x + 4\,y = -\,64$$

Problem 2

$$9\,x - 6\,y = 72$$

$$-\,4\,x + 3\,y = -\,34$$

Solve for x and y in each system using substitution.

Problem 3

$$-5x - 3y = 36$$

$$5x + y = -22$$

Problem 4

$$-7x - 5y = -6$$

$$-5x - 6y = 3$$

Solve for x and y in each system using substitution.

Problem 5

$$8x + y = 55$$

$$-8x - 2y = -54$$

Problem 6

$$-4x + 9y = -99$$

$$-9x - 9y = -18$$

Solve for x and y in each system using substitution.

Problem 7

$$x - 9y = -41$$

$$2x + 8y = 48$$

Problem 8

$$4x - 7y = -44$$

$$x + 5y = 16$$

Solve for x and y in each system using substitution.

Problem 9

$$9x - 7y = -85$$

$$8x - 5y = -67$$

Problem 10

$$5x - y = 31$$

$$-6x + 6y = -18$$

Solve for x and y in each system using substitution.

Problem 11

$$-7x - 7y = -84$$

$$-2x - 9y = -73$$

Problem 12

$$5x - 2y = -26$$

$$-7x - 8y = 4$$

Solve for x and y in each system using substitution.

Problem 13

$$3x - 2y = 8$$

$$6x + 7y = 104$$

Problem 14

$$x - 4y = 28$$

$$-3x + 9y = -57$$

Solve for x and y in each system using substitution.

Problem 15

$$8x - 6y = -106$$

$$-6x - 8y = -8$$

Problem 16

$$6x - 6y = -6$$

$$-9x + 5y = 41$$

Solve for x and y in each system using substitution.

Problem 17

$$-2x + 7y = -17$$

$$6x - 6y = 6$$

Problem 18

$$-6x - 6y = -6$$

$$-6x - 7y = -13$$

Solve for x and y in each system using substitution.

Problem 19

$$2x + 3y = 15$$

$$7x + 9y = 39$$

Problem 20

$$-4x + 6y = -2$$

$$9x - 7y = 37$$

Solve for x and y in each system using substitution.

Problem 21

$$4x - 4y = 60$$

$$2x - 3y = 39$$

Problem 22

$$5x - 3y = 2$$

$$-5x - 5y = -10$$

Solve for x and y in each system using substitution.

Problem 23

$$3x + 4y = 24$$

$$7x + 2y = 34$$

Problem 24

$$5x + 4y = 22$$

$$8x + 4y = 40$$

Solve for x and y in each system using substitution.

Problem 25

$$-6x - 5y = -19$$

$$-4x + 7y = 39$$

Problem 26

$$8x + 8y = 32$$

$$8x - 4y = 56$$

Solve for x and y in each system using substitution.

Problem 27

$$9x + 9y = 135$$

$$-8x - y = -64$$

Problem 28

$$-8x - 7y = -27$$

$$3x + 8y = -6$$

Solve for x and y in each system using substitution.

Problem 29

$3\,x + 8\,y = -\,51$

$2\,x + 9\,y = -\,67$

Problem 30

$-\,4\,x + 2\,y = -\,48$

$-\,2\,x - 9\,y = 36$

Chapter 2

Substitution with 3 Unknowns

Concepts

Three equations with three unknowns can be solved by making two substitutions.

First, isolate one unknown and plug the result into the two other equations.

This results in two equations with two unknowns, which can be solved with the method from Chapter 1.

Consider the system of three equations below.

$$2x - 3y + 4z = 4$$
$$3x + 2y + 3z = 7$$
$$5x + 8y - 2z = -7$$

Isolating x in the first equation and plugging that expression into the other two equations results in two equations with two unknowns (y and z). These two equations can then be solved using the technique from Chapter 1.

Strategy in Words

Step 1: First, isolate one unknown (x, y, or z) in one equation.

Step 2: Then substitute the expression for that unknown into the two unused equations.

> **Tip**: Be sure to plug the expression into both unused equations, but not the equation that you already used.

Step 3: Now isolate a different unknown in one of the two new equations (not the equation that you first used).

Step 4: Substitute this expression into the last equation.

> **Tip**: The last equation is the one where you have not yet isolated an unknown.

Step 5: You should now have one equation with just one unknown. Isolate that unknown to solve for it.

Step 6: Once you solve for one unknown, plug it back into any equation to find the other unknowns.

Guided Example

Solve the following system for x, y, and z.

$$2\,x-3\,y+4\,z=4$$
$$3\,x+2\,y+3\,z=7$$
$$5\,x+8\,y-2\,z=-7$$

Step 1: Isolate x in the first equation.

Do this by adding $3\,y-4\,z$ to both sides, and then dividing both sides by 2.

$$2\,x=3\,y-4\,z+4$$
$$x=\frac{3\,y}{2}-\frac{4\,z}{2}+2$$
$$x=\frac{3\,y}{2}-2\,z+2$$

(Note that $\frac{4\,z}{2}$ reduces to $2\,z$.)

Step 2: Plug this expression in for x in the bottom two equations.

$$3\left(\frac{3\,y}{2}-2\,z+2\right)+2\,y+3\,z=7$$
$$5\left(\frac{3\,y}{2}-2\,z+2\right)+8\,y-2\,z=-7$$

These two equations came from the top of the page

Simplify these equations. First distribute the coefficient.

$$\frac{9\,y}{2} - 6\,z + 6 + 2\,y + 3\,z = 7$$
$$\frac{15y}{2} - 10\,z + 10 + 8\,y - 2\,z = -\,7$$

Now group like terms together.

$$\frac{9\,y}{2} + 2\,y - 6\,z + 3\,z = 7 - 6$$
$$\frac{15\,y}{2} + 8\,y - 10\,z - 2\,z = -\,7 - 10$$

Combine like terms. When one term is a fraction, this requires first finding a common denominator.

$$\frac{13\,y}{2} - 3\,z = 1$$
$$\frac{31\,y}{2} - 12\,z = -\,17$$

Note that

$$\frac{9\,y}{2} + 2\,y = \frac{9\,y}{2} + \frac{4\,y}{2} = \frac{13\,y}{2}$$

and

$$\frac{15\,y}{2} + 8\,y = \frac{15\,y}{2} + \frac{16\,y}{2} = \frac{31\,y}{2}$$

Step 3: Isolate y in the first expression above.

Add $3\,z$ to both sides, and then multiply by $\frac{2}{13}$.

$$\frac{13\,y}{2} = 3\,z + 1$$
$$y = \frac{6\,z}{13} + \frac{2}{13}$$

Step 4: Plug this expression for y into the unused equation (the final equation from Step 2 has not been used).

$$\frac{31\,y}{2} - 12\,z = -17$$

$$\frac{31}{2}\left(\frac{6\,z}{13} + \frac{2}{13}\right) - 12\,z = -17$$

Step 5: Solve for z in this equation.

First, distribute the 31.

$$\frac{1}{2}\left(\frac{186\,z}{13} + \frac{62}{13}\right) - 12\,z = -17$$

Now distribute the $\frac{1}{2}$.

$$\frac{93\,z}{13} + \frac{31}{13} - 12\,z = -17$$

Group like terms together.

$$\frac{93\,z}{13} - 12\,z = -17 - \frac{31}{13}$$

Find a common denominator to combine like terms.

For example,

$$\frac{93\,z}{13} - 12\,z = \frac{93\,z}{13} - \frac{156\,z}{13} = -\frac{63\,z}{13}$$

$$-\frac{63\,z}{13} = -\frac{252}{13}$$

Multiply both sides by $-\frac{13}{63}$.

$$z = 4$$

Step 6: Use this value of z to solve for x and y.

Plug $z = 4$ into the equation from Step 3 with y isolated.

$$y = \frac{6\,z}{13} + \frac{2}{13}$$

$$y = \frac{6\,(4)}{13} + \frac{2}{13} = \frac{24}{13} + \frac{2}{13} = \frac{26}{13} = 2$$

Plug $y = 2$ and $z = 4$ into the equation from Step 1 with x isolated.

$$x = \frac{3\,y}{2} - 2\,z + 2$$

$$x = \frac{3\,(2)}{2} - 2\,(4) + 2 = 3 - 8 + 2 = -\,3$$

The final answers are:

$$x = -\,3,\ y = 2, \text{ and } z = 4$$

Check Your Answers

You can check your answers by plugging x, y, and z into the original equations.

$$2x - 3y + 4z = 4$$
$$3x + 2y + 3z = 7$$
$$5x + 8y - 2z = -7$$

First plug $x = -3$, $y = 2$, and $z = 4$ into the top equation.

$$2x - 3y + 4z = 4$$
$$2(-3) - 3(2) + 4(4) = 4$$
$$-6 - 6 + 16 = 4$$
$$4 = 4$$

Since $4 = 4$, we know that x, y, and z satisfy the top equation.

Now plug $x = -3$, $y = 2$, and $z = 4$ into the middle equation.

$$3x + 2y + 3z = 7$$
$$3(-3) + 2(2) + 3(4) = 7$$
$$-9 + 4 + 12 = 7$$
$$7 = 7$$

Since $7 = 7$, we know that x, y, and z satisfy the middle equation.

Finally, plug $x = -3$, $y = 2$, and $z = 4$ into the bottom equation.

$$5x + 8y - 2z = -7$$
$$5(-3) + 8(2) - 2(4) = -7$$
$$-15 + 16 - 8 = -7$$
$$-7 = -7$$

Since $-7 = -7$, we know that x, y, and z satisfy the bottom equation.

Since all three equations check out, we know that the answers for x, y, and z are correct.

Solve for x, y, and z in each system using substitution.

Problem 1

$$x + 8y + 8z = 100$$

$$7x + y - 2z = -33$$

$$-9z = -54$$

Solve for x, y, and z in each system using substitution.

Problem 2

$$-x + 5y + 5z = -35$$

$$5x + 5y + 2z = -59$$

$$7x + 8y + 6z = -95$$

Solve for x, y, and z in each system using substitution.

Problem 3

$$-6x + 2z = -24$$

$$3x - 3y - z = -12$$

$$5x + 9y + z = 76$$

Solve for x, y, and z in each system using substitution.

Problem 4

$$-9x - 3y - 4z = -65$$

$$x - 4y - 6z = 18$$

$$x + 4z = 4$$

Solve for x, y, and z in each system using substitution.

Problem 5

$$8x + 5y - 3z = -48$$

$$x - 7z = 11$$

$$7x - 9y + 3z = 27$$

Solve for x, y, and z in each system using substitution.

Problem 6

$$x - 7y - 6z = 8$$

$$-2x - 2y + 5z = 45$$

$$7x - 5y + 5z = 27$$

Solve for x, y, and z in each system using substitution.

Problem 7

$$-8x + 4y - 6z = -74$$

$$-9x + 9y + 4z = -138$$

$$-3x + 6y + 4z = -69$$

Solve for x, y, and z in each system using substitution.

Problem 8

$$-7x + 7y + 8z = 167$$

$$-6x - 2y - z = 32$$

$$-9x - 5y - 5z = 11$$

Solve for x, y, and z in each system using substitution.

Problem 9

$$-6x + 2y - 3z = -59$$

$$5x + 7y - 8z = -22$$

$$-4x + 4y + 9z = -47$$

Solve for x, y, and z in each system using substitution.

Problem 10

$$-4x - 4y = 12$$

$$-9x + 9y - 4z = 5$$

$$9x - 6y - z = -13$$

Solve for x, y, and z in each system using substitution.

Problem 11

$$-x+9y-9z=37$$

$$5x-6y-4z=-69$$

$$6y-5z=28$$

Solve for x, y, and z in each system using substitution.

Problem 12

$$3x - 6y + 3z = -30$$

$$5y - 7z = 33$$

$$6x + 9y + 8z = 110$$

Solve for x, y, and z in each system using substitution.

Problem 13

$$-6x-4y-2z=-2$$

$$-3x-5y+7z=-56$$

$$8x-y+8z=-21$$

Solve for x, y, and z in each system using substitution.

Problem 14

$$-7x - 5y - 2z = 7$$

$$5x + 9y - 9z = 58$$

$$-7x - y - z = 10$$

Solve for x, y, and z in each system using substitution.

Problem 15

$$-8x-6y+7z=-83$$

$$-7x-9y-5z=-6$$

$$8x+7y=37$$

Solve for x, y, and z in each system using substitution.

Problem 16

$$4x - 4y + 7z = -1$$

$$-5x + 2y - z = -10$$

$$-9x + 4y - 5z = -10$$

Solve for x, y, and z in each system using substitution.

Problem 17

$$6x - 7y - 4z = -11$$

$$-8x - 3y - 8z = 69$$

$$-3x + y + 6z = -17$$

Solve for x, y, and z in each system using substitution.

Problem 18

$$-2x - 6y + 4z = -10$$

$$-3x - 6y - 7z = -13$$

$$7x - 2y - 7z = -73$$

Solve for x, y, and z in each system using substitution.

Problem 19

$$-8y - 9z = 33$$

$$-x + 9y - 2z = -33$$

$$-x - 2y - 8z = 6$$

Solve for x, y, and z in each system using substitution.

Problem 20

$$4x - 2y + 6z = -22$$

$$7x + y = 44$$

$$8x - y + 3z = 19$$

Chapter 3

Simultaneous Equations

Concepts

Two equations with two unknowns can be solved simultaneously by creating equal and opposite coefficients for one of the two unknowns.

The main idea is to pick one unknown (x or y), and multiply each equation by whatever value is needed to create equal and opposite coefficients.

The reason for this is that one unknown can then be eliminated by adding the equations together.

Consider the system of equations below.

$$9x - 2y = 5$$
$$3x + 4y = 11$$

Multiplying the first equation by 2, the coefficient of y would then be -4 in the top equation and $+4$ in the bottom equation. Adding the two equations together results in one equation, with just one unknown (x).

Strategy in Words

Step 1: First, choose one unknown (x or y). Look at its coefficient in each equation. Determine the least common multiple of those two coefficients.

> A **coefficient** is a constant multiplying an unknown. For example, in $3x - 4y = 8$, the coefficient of y is -4.
>
> The least common multiple is the smallest integer that two different integers can evenly divide into. For example, the least common multiple of 6 and 8 is 24: Both 6 and 8 evenly divide into 24.

Step 2: Multiply each equation by the value needed to make each coefficient of the desired unknown equal the least common multiple.

Step 3: If the two coefficients of the desired unknown have the same sign, multiply one of the equations by -1 to create equal and opposite coefficients.

Step 4: Add the two equations together. The desired unknown will cancel in the process.

Step 5: You should now have one equation with just one unknown. Isolate that unknown to solve for it.

Step 6: Once you solve for one unknown, plug it back into any equation to find the other unknown.

Guided Example (Same Sign)

Solve the following system for x and y.

$$3\,x + 4\,y = 7$$
$$5\,x + 6\,y = 13$$

Step 1: Choose either x or y to target. Let's choose y.

The two coefficients of y are $+\,4$ and $+\,6$.

The least common multiple of 4 and 6 is 12.

Step 2: Multiply the top equation by 3 and multiply the bottom equation by 2 in order to make the coefficient of y equal 12 in each equation.

$$3\,(3\,x + 4\,y = 7)$$
$$2\,(5\,x + 6\,y = 13)$$

Multiply both sides of the equation.

$$9\,x + 12\,y = 21$$
$$10\,x + 12\,y = 26$$

Step 3: Since $12\,y$ has the same sign in both equations, multiply the bottom equation by $-\,1$ to create equal and opposite coefficients.

$$9\,x + 12\,y = 21$$
$$-\,10\,x - 12\,y = -\,26$$

Step 4: Add the two equations together. The sum of the left-hand sides equals the sum of the right-hand sides.

$$9\,x + 12\,y - 10\,x - 12\,y = 21 - 26$$

Combine like terms. The y-terms cancel out because they have equal and opposite coefficients ($+\,12$ and $-\,12$).

$$-\,x = -\,5$$

Step 5: Solve for x in this equation.

Divide both sides by $-\,1$.

$$x = 5$$

Step 4: Plug this value for x into any equation that has both x and y. One of the original equations will work.

$$3\,x + 4\,y = 7$$
$$3\,(5) + 4\,y = 7$$
$$15 + 4\,y = 7$$
$$4\,y = 7 - 15$$
$$4\,y = -\,8$$
$$y = -\,2$$

The final answers are:

$$x = 5 \text{ and } y = -\,2$$

Check Your Answers

You can check your answers by plugging x and y into the original equations.

$$3x + 4y = 7$$
$$5x + 6y = 13$$

First plug $x = 5$ and $y = -2$ into the top equation.

$$3(5) + 4(-2) = 7$$
$$15 - 8 = 7$$
$$7 = 7$$

Since $7 = 7$, we know that x and y satisfy the first equation.

Now plug $x = 5$ and $y = -2$ into the bottom equation.

$$5(5) + 6(-2) = 13$$
$$25 - 12 = 13$$
$$13 = 13$$

Since $13 = 13$, x and y also satisfy the bottom equation.

Guided Example (Opposite Sign)

Solve the following system for x and y.

$$2x - 8y = -22$$
$$5x + 3y = 37$$

Step 1: Choose either x or y to target. Let's choose y.

The two coefficients of y are -8 and $+3$.

The least common multiple of 8 and 3 is 24.

Step 2: Multiply the top equation by 3 and multiply the bottom equation by 8 in order to make the coefficient of y equal 24 in each equation.

$$3(2x - 8y = -22)$$
$$8(5x + 3y = 37)$$

Multiply both sides of the equation.

$$6x - 24y = -66$$
$$40x + 24y = 296$$

Step 3: Since $24y$ has opposite signs in the two equations, there is nothing to do in Step 3.

Step 4: Add the two equations together. The sum of the left-hand sides equals the sum of the right-hand sides.

$$6\,x - 24\,y + 40\,x + 24\,y = -\,66 + 296$$

Combine like terms. The y-terms cancel out because they have equal and opposite coefficients ($+\,24$ and $-\,24$).

$$46\,x = 230$$

Step 5: Solve for x in this equation.

Divide both sides by 46.

$$x = 5$$

Step 4: Plug this value for x into any equation that has both x and y. One of the original equations will work.

$$2\,x - 8\,y = -\,22$$
$$2\,(5) - 8\,y = -\,22$$
$$10 - 8\,y = -\,22$$
$$-\,8\,y = -\,22 - 10$$
$$-\,8\,y = -\,32$$
$$y = 4$$

The final answers are:

$x = 5$ and $y = 4$

Check Your Answers

You can check your answers by plugging x and y into the original equations.

$$2x - 8y = -22$$
$$5x + 3y = 37$$

First plug $x = 5$ and $y = 4$ into the top equation.

$$2(5) - 8(4) = -22$$
$$10 - 32 = -22$$
$$-22 = -22$$

Since $-22 = -22$, we know that x and y satisfy the first equation.

Now plug $x = 5$ and $y = 4$ into the bottom equation.

$$5(5) + 3(4) = 37$$
$$25 + 12 = 37$$
$$37 = 37$$

Since $37 = 37$, x and y also satisfy the bottom equation.

Solve for x and y by setting up simultaneous equations.

Problem 1	Problem 2
$4\,x + 2\,y = -\,14$	$3\,x - 8\,y = 87$
$8\,x + 2\,y = -\,22$	$8\,x - 5\,y = 85$

Solve for x and y by setting up simultaneous equations.

Problem 3

$$-8x + 9y = -31$$

$$-8x - 6y = 74$$

Problem 4

$$9x + 3y = -18$$

$$-9x + y = 30$$

Solve for x and y by setting up simultaneous equations.

Problem 5

$$-4x + 9y = -15$$

$$-7x - 4y = 33$$

Problem 6

$$-6x - 7y = 70$$

$$-8x + 2y = 48$$

Solve for x and y by setting up simultaneous equations.

Problem 7

$$7x - 9y = -77$$

$$9x - 9y = -81$$

Problem 8

$$4x + 6y = -76$$

$$8x + 6y = -104$$

Solve for x and y by setting up simultaneous equations.

Problem 9

$$9x - 6y = 18$$

$$-5x + 5y = 0$$

Problem 10

$$-6x + 8y = 22$$

$$-6x + 4y = 26$$

Solve for x and y by setting up simultaneous equations.

Problem 11

$$-9x - 4y = -45$$

$$-8x - 3y = -35$$

Problem 12

$$-8x - y = -56$$

$$6x + 7y = -8$$

Solve for x and y by setting up simultaneous equations.

Problem 13

$$-3x + 9y = 51$$

$$5x + 6y = -1$$

Problem 14

$$5x - 2y = -26$$

$$-2x - 3y = 37$$

Solve for x and y by setting up simultaneous equations.

Problem 15

$$7x + 8y = 9$$

$$-2x + y = 27$$

Problem 16

$$7x - 7y = -28$$

$$-x - 6y = -31$$

Solve for x and y by setting up simultaneous equations.

Problem 17

$$4x - 9y = -5$$

$$x - 7y = -6$$

Problem 18

$$7x + 4y = 8$$

$$8x - 7y = -95$$

Solve for x and y by setting up simultaneous equations.

Problem 19

$$-6x + 2y = -56$$

$$-8x - 8y = -64$$

Problem 20

$$-5x + 9y = -27$$

$$5x + 2y = 49$$

Solve for x and y by setting up simultaneous equations.

Problem 21

$$x + y = 3$$

$$6x - 4y = -72$$

Problem 22

$$4x + 4y = -28$$

$$-9x - 3y = 33$$

Solve for x and y by setting up simultaneous equations.

Problem 23

$$2x + 6y = -16$$

$$4x - 4y = 32$$

Problem 24

$$-2x + 6y = 40$$

$$6x + 6y = 72$$

Solve for x and y by setting up simultaneous equations.

Problem 25

$$-6x + 9y = 51$$

$$-4x - 9y = 19$$

Problem 26

$$6x - 9y = 81$$

$$3x + 6y = -12$$

Solve for x and y by setting up simultaneous equations.

Problem 27

$$-4x - 7y = 30$$

$$2x - 8y = 54$$

Problem 28

$$-6x - 5y = -72$$

$$-4x - 4y = -52$$

Solve for x and y by setting up simultaneous equations.

Problem 29

$$x - 4y = 21$$

$$3x + 2y = -7$$

Problem 30

$$-2x - 2y = 20$$

$$-3x - 6y = 45$$

Chapter 4

2 × 2 Determinants

Concepts

The method of Chapter 5, called Cramer's rule, requires knowing how to find the determinant of a 2×2 matrix.

An array of numbers forms a **matrix**.

The **determinant** of a 2×2 matrix is a single number defined in terms of the two diagonals as follows.

$$\det\begin{pmatrix} a & b \\ c & d \end{pmatrix} = \begin{vmatrix} a & b \\ c & d \end{vmatrix} = ad - bc$$

Multiply down to the right (ad), down to the left (bc), and then subtract bc from ad.[1]

Notation: Parentheses ( ) around an array represent a matrix, whereas straight lines around an array | | represent a determinant.

---

[1] You could multiply up to the right instead of down to the left since multiplication is commutative. That is, $bc = cb$. However, if you ever get into advanced mathematics where differential operators appear in the top row, then you must work down to the left instead of up to the right. Good habits now could pay dividends down the road.

Strategy in Words

Step 1: Multiply the top left number with the bottom right number.

Step 2: Multiply the top right number with the bottom left number.

Step 3: Subtract the result of Step 2 from the result of Step 1.

The answer is a single number.

$$\det\begin{pmatrix} a & b \\ c & d \end{pmatrix} = \begin{vmatrix} a & b \\ c & d \end{vmatrix}$$

$$\begin{vmatrix} a & b \\ c & d \end{vmatrix} = ad - bc$$

Guided Example

Find the determinant of the following matrix.

$$\begin{pmatrix} 2 & 3 \\ 4 & 5 \end{pmatrix}$$

Step 1: Multiply down to the right to make 2 times 5.

Step 2: Multiply down to the left to make 3 times 4.

Step 3: Subtract Step 2 from Step 1.

$$\det\begin{pmatrix} 2 & 3 \\ 4 & 5 \end{pmatrix} = \begin{vmatrix} 2 & 3 \\ 4 & 5 \end{vmatrix}$$

$$\det\begin{pmatrix} 2 & 3 \\ 4 & 5 \end{pmatrix} = 2(5) - 3(4)$$

$$\det\begin{pmatrix} 2 & 3 \\ 4 & 5 \end{pmatrix} = 10 - 12$$

$$\det\begin{pmatrix} 2 & 3 \\ 4 & 5 \end{pmatrix} = -2$$

The final answer is -2.

Find the determinant of each matrix.

Problem 1

$$\begin{pmatrix} 7 & 8 \\ 4 & 5 \end{pmatrix}$$

Problem 2

$$\begin{pmatrix} 2 & 6 \\ 2 & 4 \end{pmatrix}$$

Problem 3

$$\begin{pmatrix} 6 & 6 \\ 2 & 6 \end{pmatrix}$$

Problem 4

$$\begin{pmatrix} 6 & 7 \\ 0 & 0 \end{pmatrix}$$

Find the determinant of each matrix.

Problem 5

$$\begin{pmatrix} 4 & 0 \\ 1 & 8 \end{pmatrix}$$

Problem 6

$$\begin{pmatrix} 2 & 8 \\ 9 & 2 \end{pmatrix}$$

Problem 7

$$\begin{pmatrix} 6 & 0 \\ 3 & 2 \end{pmatrix}$$

Problem 8

$$\begin{pmatrix} 6 & 7 \\ 9 & 4 \end{pmatrix}$$

Find the determinant of each matrix.

Problem 9

$$\begin{pmatrix} 4 & 2 \\ 6 & 6 \end{pmatrix}$$

Problem 10

$$\begin{pmatrix} 3 & 8 \\ 4 & 0 \end{pmatrix}$$

Problem 11

$$\begin{pmatrix} 2 & 0 \\ 4 & 7 \end{pmatrix}$$

Problem 12

$$\begin{pmatrix} 0 & 0 \\ 7 & 0 \end{pmatrix}$$

Find the determinant of each matrix.

Problem 13

$$\begin{pmatrix} 0 & 4 \\ 2 & 7 \end{pmatrix}$$

Problem 14

$$\begin{pmatrix} 3 & 6 \\ 7 & 7 \end{pmatrix}$$

Problem 15

$$\begin{pmatrix} 9 & 4 \\ 5 & 0 \end{pmatrix}$$

Problem 16

$$\begin{pmatrix} 9 & 0 \\ 9 & 0 \end{pmatrix}$$

Find the determinant of each matrix.

Problem 17

$$\begin{pmatrix} 6 & -7 \\ 5 & 8 \end{pmatrix}$$

Problem 18

$$\begin{pmatrix} -9 & 3 \\ 1 & -6 \end{pmatrix}$$

Problem 19

$$\begin{pmatrix} -3 & 3 \\ -8 & -5 \end{pmatrix}$$

Problem 20

$$\begin{pmatrix} 8 & 7 \\ 6 & -6 \end{pmatrix}$$

Find the determinant of each matrix.

Problem 21

$$\begin{pmatrix} -6 & -8 \\ -3 & -5 \end{pmatrix}$$

Problem 22

$$\begin{pmatrix} -9 & 3 \\ -9 & 2 \end{pmatrix}$$

Problem 23

$$\begin{pmatrix} -4 & 4 \\ -1 & 5 \end{pmatrix}$$

Problem 24

$$\begin{pmatrix} -2 & -4 \\ -2 & -3 \end{pmatrix}$$

Find the determinant of each matrix.

Problem 25

$$\begin{pmatrix} 7 & -6 \\ -5 & -4 \end{pmatrix}$$

Problem 26

$$\begin{pmatrix} -7 & -8 \\ 2 & -4 \end{pmatrix}$$

Problem 27

$$\begin{pmatrix} 0 & -7 \\ -2 & -9 \end{pmatrix}$$

Problem 28

$$\begin{pmatrix} -6 & -2 \\ 6 & -4 \end{pmatrix}$$

Find the determinant of each matrix.

Problem 29

$$\begin{pmatrix} -6 & 3 \\ 7 & -8 \end{pmatrix}$$

Problem 30

$$\begin{pmatrix} -6 & -9 \\ 1 & 7 \end{pmatrix}$$

Problem 31

$$\begin{pmatrix} -8 & -6 \\ 4 & -4 \end{pmatrix}$$

Problem 32

$$\begin{pmatrix} 4 & -2 \\ 5 & -3 \end{pmatrix}$$

Find the determinant of each matrix.

Problem 33

$$\begin{pmatrix} -4 & -6 \\ -9 & 3 \end{pmatrix}$$

Problem 34

$$\begin{pmatrix} 9 & 3 \\ -7 & 0 \end{pmatrix}$$

Problem 35

$$\begin{pmatrix} -2 & 9 \\ 6 & -3 \end{pmatrix}$$

Problem 36

$$\begin{pmatrix} 8 & 8 \\ -8 & 5 \end{pmatrix}$$

Find the determinant of each matrix.

Problem 37

$$\begin{pmatrix} -6 & -1 \\ 0 & -1 \end{pmatrix}$$

Problem 38

$$\begin{pmatrix} 0 & 0 \\ 4 & -7 \end{pmatrix}$$

Problem 39

$$\begin{pmatrix} -6 & -4 \\ -6 & -7 \end{pmatrix}$$

Problem 40

$$\begin{pmatrix} -7 & -2 \\ -2 & -2 \end{pmatrix}$$

Chapter 5

Cramer's Rule

Concepts

Cramer's rule uses determinants to solve a system of equations.

Consider the system of two equations below.

$$a_1x + b_1y = c_1$$
$$a_2x + b_2y = c_2$$

Here, x and y are the unknowns (called the **variables**), while a_1, a_2, b_1, b_2, c_1, and c_2 are the **constants**. Specifically, a_1 and a_2 are the coefficients of x, while b_1 and b_2 are the coefficients of y.

Cramer's rule uses three 2×2 determinants to solve for x and y. Each determinant consists of four of the constants, as outlined in the strategy that follows.

Strategy in Words

Step 1: First, identify the values of a_1, a_2, b_1, b_2, c_1, and c_2.

a_1 and a_2 are the coefficients of x.
b_1 and b_2 are the coefficients of y.
c_1 and c_2 are the constants on the right-hand side.

Step 2: Make the following determinant, which uses c_1 and c_2 in place of the coefficients of x. Call this determinant D_x.

$$D_x = \begin{vmatrix} c_1 & b_1 \\ c_2 & b_2 \end{vmatrix}$$

Step 3: Make the following determinant, which uses c_1 and c_2 in place of the coefficients of y. Call this determinant D_y.

$$D_y = \begin{vmatrix} a_1 & c_1 \\ a_2 & c_2 \end{vmatrix}$$

Step 4: Make the following determinant, which only uses coefficients. Call this determinant D_c.

$$D_c = \begin{vmatrix} a_1 & b_1 \\ a_2 & b_2 \end{vmatrix}$$

Steps 5 and 6: Find x and y by dividing D_x and D_y by D_c.

$$x = \frac{D_x}{D_c} \quad , \quad y = \frac{D_y}{D_c}$$

Guided Example

Solve the following system for x and y.

$$2x - 3y = -6$$
$$4x + y = 16$$

Step 1: Identify the values of a_1, a_2, b_1, b_2, c_1, and c_2.

$$a_1 = 2$$
$$a_2 = 4$$
$$b_1 = -3$$
$$b_2 = 1$$
$$c_1 = -6$$
$$c_2 = 16$$

Note that the coefficient of $+y$ equals 1.

Step 2: Find the determinant D_x.

$$D_x = \begin{vmatrix} c_1 & b_1 \\ c_2 & b_2 \end{vmatrix} = \begin{vmatrix} -6 & -3 \\ 16 & 1 \end{vmatrix}$$
$$D_x = -6(1) - (-3)(16)$$
$$D_x = -6 + 48$$
$$D_x = 42$$

Step 3: Find the determinant D_y.

$$D_y = \begin{vmatrix} a_1 & c_1 \\ a_2 & c_2 \end{vmatrix} = \begin{vmatrix} 2 & -6 \\ 4 & 16 \end{vmatrix}$$
$$D_y = 2(16) - (-6)(4)$$

$$D_y = 32 + 24$$
$$D_y = 56$$

Step 4: Find the determinant D_c.

$$D_c = \begin{vmatrix} a_1 & b_1 \\ a_2 & b_2 \end{vmatrix} = \begin{vmatrix} 2 & -3 \\ 4 & 1 \end{vmatrix}$$
$$D_c = 2\,(1) - (-3)\,(4)$$
$$D_c = 2 + 12$$
$$D_c = 14$$

Step 5: Calculate x.

$$x = \frac{D_x}{D_c} = \frac{42}{14} = 3$$

Step 6: Calculate y.

$$y = \frac{D_y}{D_c} = \frac{56}{14} = 4$$

The final answers are:

$$x = 3 \text{ and } y = 4$$

Check Your Answers

You can check your answers by plugging x and y into the original equations.

$$2x - 3y = -6$$
$$4x + y = 16$$

First plug $x = 3$ and $y = 4$ into the top equation.

$$2(3) - 3(4) = -6$$
$$6 - 12 = -6$$
$$-6 = -6$$

Since $-6 = -6$, we know that x and y satisfy the first equation.

Now plug $x = 3$ and $y = 4$ into the bottom equation.

$$4(3) + 4 = 16$$
$$12 + 4 = 16$$
$$16 = 16$$

Since $16 = 16$, x and y also satisfy the bottom equation.

Use Cramer's rule to solve for x and y in each system.

Problem 1

$$6x + 2y = 48$$

$$5x + 2y = 42$$

Problem 2

$$5x - y = -41$$

$$6x + y = -47$$

Use Cramer's rule to solve for x and y in each system.

Problem 3

$$-2x + y = -1$$

$$-9x + 8y = -36$$

Problem 4

$$-7x - 8y = -44$$

$$2x - 6y = -62$$

Use Cramer's rule to solve for x and y in each system.

Problem 5

$$-9x + 9y = -108$$

$$x + 3y = -4$$

Problem 6

$$-5x + 3y = 1$$

$$-3x - 3y = -33$$

Use Cramer's rule to solve for x and y in each system.

Problem 7

$$-6x - 2y = 4$$

$$-2x + 5y = 41$$

Problem 8

$$-5x - 9y = 35$$

$$2x - 8y = 44$$

Use Cramer's rule to solve for x and y in each system.

Problem 9

$$-6x - 7y = 16$$

$$2x - 5y = -20$$

Problem 10

$$3x + y = -2$$

$$-9x + 6y = 15$$

Use Cramer's rule to solve for x and y in each system.

Problem 11

$$2x + 2y = -4$$

$$-x - 8y = -19$$

Problem 12

$$-5x + 4y = -14$$

$$5x + 5y = 50$$

Use Cramer's rule to solve for x and y in each system.

Problem 13

$$5\,x + 3\,y = 27$$

$$6\,x - 9\,y = 108$$

Problem 14

$$3\,x - 7\,y = -\,48$$

$$x + 3\,y = 0$$

Use Cramer's rule to solve for x and y in each system.

Problem 15

$$-3x + 8y = 51$$

$$5x - 2y = 17$$

Problem 16

$$x + y = -12$$

$$-6x + 6y = 72$$

Use Cramer's rule to solve for x and y in each system.

Problem 17

$$3x + 6y = -9$$

$$4x + 7y = -11$$

Problem 18

$$3x + 9y = -6$$

$$-9x + 3y = 78$$

Use Cramer's rule to solve for x and y in each system.

Problem 19

$$3x - 9y = -96$$

$$-3x - 3y = 0$$

Problem 20

$$3x + 3y = -15$$

$$-7x + 7y = -35$$

Use Cramer's rule to solve for x and y in each system.

Problem 21

$$4x - 2y = -4$$

$$8x + 4y = -24$$

Problem 22

$$-x - 9y = -26$$

$$-7x - 4y = -64$$

Use Cramer's rule to solve for x and y in each system.

Problem 23

$$9x + 2y = -24$$

$$-2x - 2y = 10$$

Problem 24

$$-6x + 5y = 37$$

$$-4x + 2y = 18$$

Use Cramer's rule to solve for x and y in each system.

Problem 25

$$-7x - y = -31$$

$$4x - 5y = 1$$

Problem 26

$$8x - 5y = -78$$

$$-5x + 8y = 78$$

Use Cramer's rule to solve for x and y in each system.

Problem 27

$$-4x + 3y = -24$$

$$-8x + 5y = -52$$

Problem 28

$$-3x - 7y = -33$$

$$-7x - 6y = -46$$

Use Cramer's rule to solve for x and y in each system.

Problem 29

$$4x + 6y = -14$$

$$-6x - 2y = -14$$

Problem 30

$$7x - 5y = 44$$

$$9x - 8y = 55$$

Chapter 6

3 × 3 Determinants

Concepts

There are two ways to find the determinant of a 3×3 matrix.

One way is called the method of cofactors. Each element along the top row is used as a **cofactor**. Block out the row and column of a cofactor to determine its **submatrix**.

For example, in the matrix below, the row and column of a particular cofactor (a) are blocked out, leaving a 2×2 submatrix.

$$\begin{pmatrix} a & b & c \\ d & e & f \\ g & h & i \end{pmatrix}$$

To find the determinant of a 3×3 matrix using the method of cofactors, multiply each cofactor by the determinant of its corresponding submatrix, alternating sign as follows.

$$\begin{vmatrix} a & b & c \\ d & e & f \\ g & h & i \end{vmatrix} = a\begin{vmatrix} e & f \\ h & i \end{vmatrix} - b\begin{vmatrix} d & f \\ g & i \end{vmatrix} + c\begin{vmatrix} d & e \\ g & h \end{vmatrix}$$

Evaluating the 2 × 2 determinants, this simplifies to:

$$\begin{vmatrix} a & b & c \\ d & e & f \\ g & h & i \end{vmatrix} = aei - afh - bdi + bfg + cdh - ceg$$

There is also a shortcut method for finding the determinant of a 3 × 3 matrix. For the shortcut method, rewrite the matrix repeating the left two columns at the right, as shown below.

$$\begin{array}{ccc:cc} a & b & c & a & b \\ d & e & f & d & e \\ g & h & i & g & h \end{array}$$

Now multiply down to the right across the three diagonals, multiply down to the left across three more diagonals, and subtract the latter three terms from the first three terms, as shown below.

$$\begin{array}{ccccc} a & b & c & a & b \\ d & e & f & d & e \\ g & h & i & g & h \end{array}$$

$$\begin{vmatrix} a & b & c \\ d & e & f \\ g & h & i \end{vmatrix} = aei + bfg + cdh - ceg - afh - bdi$$

Observe that the shortcut method yields the same result as the cofactor method.[2]

---

[2] While the shortcut method is more convenient, only the cofactor method generalizes to 4 × 4 determinants.

Strategy in Words

Step 1: First repeat the left two columns to the right of the given matrix.

Step 2: Multiply down to the right along the three diagonals.

Step 3: Multiply down to the left along the three diagonals.

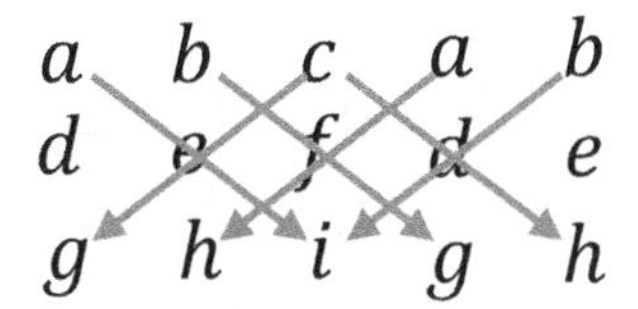

Step 3: Subtract the three terms of Step 3 from the three terms of Step 2.

The answer is a single number.

$$\begin{vmatrix} a & b & c \\ d & e & f \\ g & h & i \end{vmatrix} = aei + bfg + cdh - ceg - afh - bdi$$

Guided Example

Find the determinant of the following matrix.

$$\begin{pmatrix} 3 & 1 & 4 \\ 2 & 5 & 6 \\ 4 & 3 & 2 \end{pmatrix}$$

Step 1: Copy the left two columns and join them to the right.

$$\begin{matrix} 3 & 1 & 4 & 3 & 1 \\ 2 & 5 & 6 & 2 & 5 \\ 4 & 3 & 2 & 4 & 3 \end{matrix}$$

Step 2: Multiply down to the right along the three diagonals.

$$3\,(5)\,(2)$$
$$1\,(6)\,(4)$$
$$4\,(2)\,(3)$$

Step 3: Multiply down to the left along the three diagonals.

$$4\,(5)\,(4)$$
$$3\,(6)\,(3)$$
$$1\,(2)\,(2)$$

Step 4: Subtract the three terms of Step 3 from the three terms of Step 2.

$$3(5)(2) + 1(6)(4) + 4(2)(3) - 4(5)(4) - 3(6)(3) - 1(2)(2)$$

$$\begin{vmatrix} 3 & 1 & 4 \\ 2 & 5 & 6 \\ 4 & 3 & 2 \end{vmatrix} = 30 + 24 + 24 - 80 - 54 - 4$$

$$\begin{vmatrix} 3 & 1 & 4 \\ 2 & 5 & 6 \\ 4 & 3 & 2 \end{vmatrix} = 78 - 138$$

$$\begin{vmatrix} 3 & 1 & 4 \\ 2 & 5 & 6 \\ 4 & 3 & 2 \end{vmatrix} = -60$$

The final answer is -60.

Find the determinant of each matrix.

Problem 1

$$\begin{pmatrix} 0 & 2 & 8 \\ 8 & 1 & 4 \\ 6 & 8 & 4 \end{pmatrix}$$

Problem 2

$$\begin{pmatrix} 6 & 1 & 4 \\ 0 & 2 & 9 \\ 1 & 2 & 3 \end{pmatrix}$$

Find the determinant of each matrix.

Problem 3

$$\begin{pmatrix} 2 & 2 & 6 \\ 3 & 1 & 1 \\ 0 & 5 & 5 \end{pmatrix}$$

Problem 4

$$\begin{pmatrix} 2 & 9 & 6 \\ 1 & 0 & 8 \\ 9 & 6 & 5 \end{pmatrix}$$

Find the determinant of each matrix.

Problem 5

$$\begin{pmatrix} 4 & 9 & 8 \\ 8 & 8 & 4 \\ 3 & 8 & 1 \end{pmatrix}$$

Problem 6

$$\begin{pmatrix} 6 & 5 & 7 \\ 6 & 1 & 1 \\ 9 & 2 & 2 \end{pmatrix}$$

Find the determinant of each matrix.

Problem 7

$$\begin{pmatrix} 5 & 2 & 2 \\ 5 & 1 & 5 \\ 6 & 2 & 7 \end{pmatrix}$$

Problem 8

$$\begin{pmatrix} 7 & 7 & 2 \\ 7 & 8 & 1 \\ 0 & 8 & 0 \end{pmatrix}$$

Find the determinant of each matrix.

Problem 9

$$\begin{pmatrix} -3 & 6 & 3 \\ -7 & -9 & 8 \\ -6 & -3 & -1 \end{pmatrix}$$

Problem 10

$$\begin{pmatrix} -1 & -8 & 5 \\ -4 & -9 & 5 \\ 6 & 1 & 4 \end{pmatrix}$$

Find the determinant of each matrix.

Problem 11

$$\begin{pmatrix} 1 & 9 & 6 \\ -3 & 7 & 9 \\ -1 & 7 & 5 \end{pmatrix}$$

Problem 12

$$\begin{pmatrix} -4 & -9 & -4 \\ -6 & -3 & -3 \\ 6 & -4 & -2 \end{pmatrix}$$

Find the determinant of each matrix.

Problem 13

$$\begin{pmatrix} -2 & -3 & -7 \\ 8 & 6 & 7 \\ -5 & 9 & 8 \end{pmatrix}$$

Problem 14

$$\begin{pmatrix} 0 & -6 & -2 \\ 6 & 1 & 4 \\ 3 & -7 & -4 \end{pmatrix}$$

Find the determinant of each matrix.

Problem 15

$$\begin{pmatrix} 8 & -7 & -8 \\ -5 & 6 & 2 \\ -1 & 5 & 3 \end{pmatrix}$$

Problem 16

$$\begin{pmatrix} 6 & 7 & -2 \\ 0 & -8 & 5 \\ 7 & -9 & -8 \end{pmatrix}$$

Find the determinant of each matrix.

Problem 17

$$\begin{pmatrix} 6 & 5 & 3 \\ -2 & 2 & 5 \\ 5 & 1 & 3 \end{pmatrix}$$

Problem 18

$$\begin{pmatrix} 3 & 2 & -9 \\ -5 & -3 & 9 \\ 0 & 8 & 7 \end{pmatrix}$$

Find the determinant of each matrix.

Problem 19

$$\begin{pmatrix} -8 & 1 & 9 \\ -3 & -2 & 2 \\ -1 & -6 & -4 \end{pmatrix}$$

Problem 20

$$\begin{pmatrix} -4 & 5 & -3 \\ 7 & 8 & 5 \\ 8 & -9 & -4 \end{pmatrix}$$

Chapter 7

Cramer's Rule with 3 Unknowns

Concepts

A system of three equations with three unknowns can be solved using Cramer's rule with 3×3 determinants.

Consider the system of three equations below.

$$a_1x + b_1y + c_1z = d_1$$
$$a_2x + b_2y + c_2z = d_2$$
$$a_3x + b_3y + c_3z = d_3$$

Here, a_1, a_2, and a_3 are the coefficients of x; b_1, b_2, and b_3 are the coefficients of y; c_1, c_2, and c_3 are the coefficients of z; and d_1, d_2, and d_3 are the constants on the right-hand side.

Strategy in Words

Step 1: First, identify the values of the constants.

a_1, a_2, and a_3 are the coefficients of x.
b_1, b_2, and b_3 are the coefficients of y.
c_1, c_2, and c_3 are the coefficients of z.
d_1, d_2, and d_3 are the constants on the right-hand side.

Step 2: Make the following determinant, which uses d_1, d_2, and d_3 in place of the coefficients of x. Call this determinant D_x.

$$D_x = \begin{vmatrix} d_1 & b_1 & c_1 \\ d_2 & b_2 & c_2 \\ d_3 & b_3 & c_3 \end{vmatrix}$$

Step 3: Make the following determinant, which uses d_1, d_2, and d_3 in place of the coefficients of y. Call this determinant D_y.

$$D_y = \begin{vmatrix} a_1 & d_1 & c_1 \\ a_2 & d_2 & c_2 \\ a_3 & d_3 & c_3 \end{vmatrix}$$

Step 4: Make the following determinant, which uses d_1, d_2, and d_3 in place of the coefficients of z. Call this determinant D_z.

$$D_z = \begin{vmatrix} a_1 & b_1 & d_1 \\ a_2 & b_2 & d_2 \\ a_3 & b_3 & d_3 \end{vmatrix}$$

Step 5: Make the following determinant, which only uses coefficients. Call this determinant D_c.

$$D_c = \begin{vmatrix} a_1 & b_1 & c_1 \\ a_2 & b_2 & c_2 \\ a_3 & b_3 & c_3 \end{vmatrix}$$

Step 6: Find x by dividing D_x by D_c.

$$x = \frac{D_x}{D_c}$$

Step 7: Find y by dividing D_y by D_c.

$$y = \frac{D_y}{D_c}$$

Step 8: Find z by dividing D_z by D_c.

$$z = \frac{D_z}{D_c}$$

Guided Example (with All Coefficients)

Solve the following system for x, y, and z.

$$4\,x - 3\,y + 5\,z = 40$$
$$2\,x + y - 3\,z = -\,8$$
$$3\,x - y - 2\,z = -\,2$$

Step 1: Identify the values of the constants.

$$a_1 = 4$$
$$a_2 = 2$$
$$a_3 = 3$$
$$b_1 = -\,3$$
$$b_2 = 1$$
$$b_3 = -\,1$$
$$c_1 = 5$$
$$c_2 = -\,3$$
$$c_3 = -\,2$$
$$d_1 = 40$$
$$d_2 = -\,8$$
$$d_3 = -\,2$$

Step 2: Find the determinant D_x.

$$D_x = \begin{vmatrix} d_1 & b_1 & c_1 \\ d_2 & b_2 & c_2 \\ d_3 & b_3 & c_3 \end{vmatrix} = \begin{vmatrix} 40 & -\,3 & 5 \\ -\,8 & 1 & -\,3 \\ -\,2 & -\,1 & -\,2 \end{vmatrix}$$

$$D_x = 40(1)(-2) + (-3)(-3)(-2) + 5(-8)(-1)$$
$$-5(1)(-2) - 40(-3)(-1) - (-3)(-8)(-2)$$
$$D_x = -80 - 18 + 40 + 10 - 120 + 48$$
$$D_x = -120$$

Step 3: Find the determinant D_y.

$$D_y = \begin{vmatrix} a_1 & d_1 & c_1 \\ a_2 & d_2 & c_2 \\ a_3 & d_3 & c_3 \end{vmatrix} = \begin{vmatrix} 4 & 40 & 5 \\ 2 & -8 & -3 \\ 3 & -2 & -2 \end{vmatrix}$$
$$D_y = 4(-8)(-2) + 40(-3)(3) + 5(2)(-2)$$
$$-5(-8)(3) - 4(-3)(-2) - 40(2)(-2)$$
$$D_y = 64 - 360 - 20 + 120 - 24 + 160$$
$$D_y = -60$$

Step 4: Find the determinant D_z.

$$D_z = \begin{vmatrix} a_1 & b_1 & d_1 \\ a_2 & b_2 & d_2 \\ a_3 & b_3 & d_3 \end{vmatrix} = \begin{vmatrix} 4 & -3 & 40 \\ 2 & 1 & -8 \\ 3 & -1 & -2 \end{vmatrix}$$
$$D_z = 4(1)(-2) + (-3)(-8)(3) + 40(2)(-1)$$
$$-40(1)(3) - 4(-8)(-1) - (-3)(2)(-2)$$
$$D_z = -8 + 72 - 80 - 120 - 32 - 12$$
$$D_z = -180$$

Step 5: Find the determinant D_c.

$$D_c = \begin{vmatrix} a_1 & b_1 & c_1 \\ a_2 & b_2 & c_2 \\ a_3 & b_3 & c_3 \end{vmatrix} = \begin{vmatrix} 4 & -3 & 5 \\ 2 & 1 & -3 \\ 3 & -1 & -2 \end{vmatrix}$$

$$D_c = 4(1)(-2) + (-3)(-3)(3) + 5(2)(-1)$$
$$-5(1)(3) - 4(-3)(-1) - (-3)(2)(-2)$$
$$D_c = -8 + 27 - 10 - 15 - 12 - 12$$
$$D_c = -30$$

Step 6: Calculate x.

$$x = \frac{D_x}{D_c} = \frac{-120}{-30} = 4$$

Step 7: Calculate y.

$$y = \frac{D_y}{D_c} = \frac{-60}{-30} = 2$$

Step 8: Calculate z.

$$z = \frac{D_z}{D_c} = \frac{-180}{-30} = 6$$

The final answers are:

$$x = 4, y = 2, \text{ and } z = 6$$

Check Your Answers

You can check your answers by plugging x, y, and z into the original equations.

$$4x - 3y + 5z = 40$$
$$2x + y - 3z = -8$$
$$3x - y - 2z = -2$$

First plug $x = 4$, $y = 2$, and $z = 6$ into the top equation.

$$4x - 3y + 5z = 40$$
$$4(4) - 3(2) + 5(6) = 40$$
$$16 - 6 + 30 = 40$$
$$40 = 40$$

Since $40 = 40$, we know that x, y, and z satisfy the top equation.

Now plug $x = 4$, $y = 2$, and $z = 6$ into the middle equation.

$$2x + y - 3z = -8$$
$$2(4) + 2 - 3(6) = -8$$
$$8 + 2 - 18 = -8$$
$$-8 = -8$$

Since $-8 = -8$, we know that x, y, and z satisfy the middle equation.

Finally, plug $x = 4$, $y = 2$, and $z = 6$ into the bottom equation.

$$3\,x - y - 2\,z = -\,2$$
$$3\,(4) - 2 - 2\,(6) = -\,2$$
$$12 - 2 - 12 = -\,2$$
$$-\,2 = -\,2$$

Since $-\,2 = -\,2$, we know that x, y, and z satisfy the bottom equation.

Since all three equations check out, we know that the answers for x, y, and z are correct.

Guided Example (with Missing Coefficients)

Solve the following system for x, y, and z.

$$2x - 4y + z = -15$$
$$3y - 2z = 14$$
$$6x + 5y = 76$$

Step 1: Identify the values of the constants.

Note: If a term is missing, its coefficient equals zero.

Since the x-term is missing in the second equation, $a_2 = 0$.

Since the z-term is missing in the third equation, $c_3 = 0$.

$$a_1 = 2$$
$$a_2 = 0$$
$$a_3 = 6$$
$$b_1 = -4$$
$$b_2 = 3$$
$$b_3 = 5$$
$$c_1 = 1$$
$$c_2 = -2$$
$$c_3 = 0$$
$$d_1 = -15$$
$$d_2 = 14$$
$$d_3 = 76$$

Step 2: Find the determinant D_x.

$$D_x = \begin{vmatrix} d_1 & b_1 & c_1 \\ d_2 & b_2 & c_2 \\ d_3 & b_3 & c_3 \end{vmatrix} = \begin{vmatrix} -15 & -4 & 1 \\ 14 & 3 & -2 \\ 76 & 5 & 0 \end{vmatrix}$$

$$D_x = (-15)(3)(0) + (-4)(-2)(76) + 1(14)(5) - 1(3)(76) - (-15)(-2)(5) - (-4)(14)(0)$$

$$D_x = 0 + 608 + 70 - 228 - 150 + 0$$

$$D_x = 300$$

Step 3: Find the determinant D_y.

$$D_y = \begin{vmatrix} a_1 & d_1 & c_1 \\ a_2 & d_2 & c_2 \\ a_3 & d_3 & c_3 \end{vmatrix} = \begin{vmatrix} 2 & -15 & 1 \\ 0 & 14 & -2 \\ 6 & 76 & 0 \end{vmatrix}$$

$$D_y = 2(14)(0) + (-15)(-2)(6) + 1(0)(76) - 1(14)(6) - 2(-2)(76) - (-15)(0)(0)$$

$$D_y = 0 + 180 + 0 - 84 + 304 + 0$$

$$D_y = 400$$

Step 4: Find the determinant D_z.

$$D_z = \begin{vmatrix} a_1 & b_1 & d_1 \\ a_2 & b_2 & d_2 \\ a_3 & b_3 & d_3 \end{vmatrix} = \begin{vmatrix} 2 & -4 & -15 \\ 0 & 3 & 14 \\ 6 & 5 & 76 \end{vmatrix}$$

$$D_z = 2(3)(76) + (-4)(14)(6) + (-15)(0)(5) - (-15)(3)(6) - 2(14)(5) - (-4)(0)(76)$$

$$D_z = 456 - 336 + 0 + 270 - 140 + 0$$

$$D_z = 250$$

Step 5: Find the determinant D_c.

$$D_c = \begin{vmatrix} a_1 & b_1 & c_1 \\ a_2 & b_2 & c_2 \\ a_3 & b_3 & c_3 \end{vmatrix} = \begin{vmatrix} 2 & -4 & 1 \\ 0 & 3 & -2 \\ 6 & 5 & 0 \end{vmatrix}$$

$$D_c = 2(3)(0) + (-4)(-2)(6) + 1(0)(5)$$
$$-1(3)(6) - 2(-2)(5) - (-4)(0)(0)$$
$$D_c = 0 + 48 + 0 - 18 + 20 + 0$$
$$D_c = 50$$

Step 6: Calculate x.

$$x = \frac{D_x}{D_c} = \frac{300}{50} = 6$$

Step 7: Calculate y.

$$y = \frac{D_y}{D_c} = \frac{400}{50} = 8$$

Step 8: Calculate z.

$$z = \frac{D_z}{D_c} = \frac{250}{50} = 5$$

The final answers are:

$$x = 6, y = 8, \text{ and } z = 5$$

Check Your Answers

You can check your answers by plugging x, y, and z into the original equations.

$$2x - 4y + z = -15$$
$$3y - 2z = 14$$
$$6x + 5y = 76$$

First plug $x = 6$, $y = 8$, and $z = 5$ into the top equation.

$$2x - 4y + z = -15$$
$$2(6) - 4(8) + 5 = -15$$
$$12 - 32 + 5 = -15$$
$$-15 = -15$$

Since $-15 = -15$, we know that x, y, and z satisfy the top equation.

Now plug $x = 6$, $y = 8$, and $z = 5$ into the middle equation.

$$3y - 2z = 14$$
$$3(8) - 2(5) = 14$$
$$24 - 10 = 14$$
$$14 = 14$$

Since $14 = 14$, we know that x, y, and z satisfy the middle equation.

Finally, plug $x = 6$, $y = 8$, and $z = 5$ into the bottom equation.

$$6x + 5y = 76$$
$$6(6) + 5(8) = 76$$
$$36 + 40 = 76$$
$$76 = 76$$

Since $76 = 76$, we know that x, y, and z satisfy the bottom equation.

Since all three equations check out, we know that the answers for x, y, and z are correct.

Use Cramer's rule to solve for x, y, and z in each system.

Problem 1

$$2\,x - 2\,z = 10$$

$$4\,x - 4\,y + 7\,z = -\,63$$

$$-\,8\,x + 5\,y + 3\,z = -\,15$$

Use Cramer's rule to solve for x, y, and z in each system.

Problem 2

$$-9x - 2y + 9z = -72$$

$$6x + 7y - 2z = 71$$

$$5x - 5y = -50$$

Use Cramer's rule to solve for x, y, and z in each system.

Problem 3

$$-9y + 4z = 82$$

$$-8x - 5y - 7z = -43$$

$$-9x - 8y + 5z = 56$$

Use Cramer's rule to solve for x, y, and z in each system.

Problem 4

$$-7x - 8y + 5z = -21$$

$$-7y + 3z = -45$$

$$9x + 4y + 7z = 51$$

Use Cramer's rule to solve for x, y, and z in each system.

Problem 5

$$-x + 2z = 14$$

$$x + 6y + 2z = 34$$

$$4x + 2y - 9z = -54$$

Use Cramer's rule to solve for x, y, and z in each system.

Problem 6

$$7x - 8y + 4z = 45$$

$$2x - 3y - 8z = -11$$

$$6y + 3z = -21$$

Use Cramer's rule to solve for x, y, and z in each system.

Problem 7

$$-7x + 5y - 4z = -37$$

$$x + 7y = -9$$

$$3x + 3y + 3z = 3$$

Use Cramer's rule to solve for x, y, and z in each system.

Problem 8

$$8\,x - 2\,y + 8\,z = 82$$

$$-\,6\,x + 3\,y + z = -\,2$$

$$3\,x - 3\,y + 5\,z = 29$$

Use Cramer's rule to solve for x, y, and z in each system.

Problem 9

$$-8x - 5y - 9z = 45$$

$$2x - 7y - 5z = -7$$

$$2x + y - 2z = 5$$

Use Cramer's rule to solve for x, y, and z in each system.

Problem 10

$$-2x + 8y - 6z = 64$$

$$-2x + 2y - 8z = 52$$

$$9x + 8z = -30$$

Use Cramer's rule to solve for x, y, and z in each system.

Problem 11

$$-6x - 9y + 6z = -33$$

$$-7x + 2y - 2z = -48$$

$$4x - 9y - 6z = -69$$

Use Cramer's rule to solve for x, y, and z in each system.

Problem 12

$$3x - 4y = -38$$

$$-3x - 8y + z = -67$$

$$3x + 8y + 2z = 40$$

Use Cramer's rule to solve for x, y, and z in each system.

Problem 13

$$2x - 7y + 3z = 45$$

$$9x - 2y - 4z = -85$$

$$2x - 8y + z = 34$$

Use Cramer's rule to solve for x, y, and z in each system.

Problem 14

$$-7x + 6y - 7z = -45$$

$$-x + y + z = -15$$

$$-4x - 8y - 2z = 12$$

Use Cramer's rule to solve for x, y, and z in each system.

Problem 15

$$2\,x - 4\,y - 2\,z = -\,16$$

$$6\,x + 7\,y + 2\,z = -\,57$$

$$6\,x + y - 3\,z = -\,69$$

Use Cramer's rule to solve for x, y, and z in each system.

Problem 16

$$-9x + 2y + 2z = 40$$

$$-7x - 2y - 9z = -57$$

$$-2y - z = -15$$

Use Cramer's rule to solve for x, y, and z in each system.

Problem 17

$$-3x + 4y + 8z = -58$$

$$-x + y + 7z = -51$$

$$8x - 4y = 32$$

Use Cramer's rule to solve for x, y, and z in each system.

Problem 18

$$-6x - 4y - 2z = -64$$

$$-9x - 2z = -61$$

$$8x - 5y + 8z = 18$$

Use Cramer's rule to solve for x, y, and z in each system.

Problem 19

$$-7y - 2z = -20$$

$$-4x - 6y + 5z = -80$$

$$9x + 4y + z = 93$$

Use Cramer's rule to solve for x, y, and z in each system.

Problem 20

$$2x - 4y - z = -2$$

$$-2x + 9y + z = -8$$

$$-x - 3y - 4z = 2$$

Appendix

Solving Systems with a TI-83 Calculator

Concepts

The goal of this workbook is to provide practice solving systems of equations by hand. However, sometimes it is handy to be able to solve a system of equations quickly when applying algebra to a field such as engineering. In that spirit, this Appendix shows how to use a popular calculator model to solve a system of equations.

These instructions are specifically designed for the Texas Instruments model TI-83 Plus, one of the more popular calculators among students taking math and science courses. Higher level TI calculators, such as the TI-89, have similar capability, though the same functions are accessed in a different way from the home screen.

The first step is to write the system of equations in **standard form**. This means to write the x-term first, followed by the y-term, and if there is a third unknown, the z-term; it also means to place the constant terms on the right-hand side. (If you need to rearrange the given equations, follow the rules of algebra when you move terms around.)

The equations should have the following structure:

$$a_1x + b_1y + c_1z = d_1$$
$$a_2x + b_2y + c_2z = d_2$$
$$a_3x + b_3y + c_3z = d_3$$

(If there are only two unknowns, there will be just two equations and no z-terms.)

Make the following 3×4 matrix with the constants.

$$\begin{pmatrix} a_1 & b_1 & c_1 & d_1 \\ a_2 & b_2 & c_2 & d_2 \\ a_3 & b_3 & c_3 & d_3 \end{pmatrix}$$

(For a system of two equations with two unknowns, instead make a 2×3 matrix with the constants.)

By entering this 3×4 matrix into a TI-83 Plus calculator, one of the built-in matrix functions can be used to solve for the unknowns, as illustrated with the following strategy.

Strategy in Words

Step 1: Write the given equations in the following form.

$$a_1x + b_1y + c_1z = d_1$$
$$a_2x + b_2y + c_2z = d_2$$
$$a_3x + b_3y + c_3z = d_3$$

Step 2: Organize the constants into the following matrix.

$$\begin{pmatrix} a_1 & b_1 & c_1 & d_1 \\ a_2 & b_2 & c_2 & d_2 \\ a_3 & b_3 & c_3 & d_3 \end{pmatrix}$$

Step 3: Visit the Matrix screen. First press the 2nd button, then find Matrix with the x^{-1} button.

Step 4: Use the arrow keys to select Edit at the top. Choose the name of a matrix, such as [A]. Press Enter.

Step 5: Change the size at the top to 3×4.

Step 6: Using the arrow keys to position the cursor, type the values for all 12 elements in the matrix. When you finish, scroll left and right and carefully check each element.

Step 7: Exit the matrix screen. Press 2nd and find Quit with the Mode button.

Step 8: Return to the matrix screen. Select Math at the top. Scroll down to find the rref function. Press Enter.

Be sure to select rref and not ref. You want the function with two r's.

Step 9: Return to the matrix screen. Under Names, select the name of your matrix and press Enter.

Step 10: Close the parentheses and press Enter.

Step 11: This step is optional. If you see decimals and would prefer fractions, press Math, select Frac, and press Enter.

Step 12: If successful, you should see a 3×4 matrix on the screen with the following form.

$$\begin{pmatrix} 1 & 0 & 0 & x \\ 0 & 1 & 0 & y \\ 0 & 0 & 1 & z \end{pmatrix}$$

The answers for x, y, and z appear in order, from top to bottom, in the rightmost column.

Note: If the left three columns don't look the same, you either made a mistake somewhere in Steps 1-11, or the system of equations isn't linearly independent (i.e. there isn't a unique solution to the system).

Calculator Example

Solve the following system for x, y, and z.

$$5x + 5z - 26 = 4y$$
$$3z - x - 9y = -7$$
$$7x - 44 + 8z = 0$$

Step 1: Write the given equations in standard form, with the x-term first, y-term next, z-term last, and the constant on the right-hand side. Use the rules of algebra to rearrange terms.

$$5x - 4y + 5z = 26$$
$$-x - 9y + 3z = -7$$
$$7x + 0y + 8z = 44$$

Step 2: Organize the constants into the prescribed 3×4 matrix.

$$\begin{pmatrix} 5 & -4 & 5 & 26 \\ -1 & -9 & 3 & -7 \\ 7 & 0 & 8 & 44 \end{pmatrix}$$

Step 3: Press the 2nd button followed by Matrix. Find Matrix with the x^{-1} button.

Step 4: Use the arrow keys to select Edit at the top. Choose matrix [A]. Press Enter.

Step 5: Change the size at the top to 3×4.

Step 6: Using the arrow keys to position the cursor, type the values for all 12 elements in the matrix. When you finish, scroll left and right and carefully check each element.

Step 7: Press 2nd followed by Quit. Find Quit with Mode.

Step 8: Press the 2nd button followed by Matrix. Select Math at the top. Scroll down to find the rref function. Press Enter.

> Be sure to select rref and not ref. You want the function with two r's.
>
> You should see the following on the main screen.

rref(

Step 9: Press the 2nd button followed by Matrix. Under Names, select matrix [A] and press Enter.

> You should see the following on the main screen.

rref([A]

Step 10: Close the parentheses and press Enter.

> Just before pressing Enter, you should see the following on the main screen.

rref([A])

Step 11: Since the answers are all integers, Step 11 doesn't apply to this example.

Step 12: If successful, the following 3×4 matrix will appear on the main screen.

$$\begin{pmatrix} 1 & 0 & 0 & 4 \\ 0 & 1 & 0 & 1 \\ 0 & 0 & 1 & 2 \end{pmatrix}$$

The answers for x, y, and z appear in order, from top to bottom, in the rightmost column.

The final answers are:

$$x = 4, y = 1, \text{and } z = 2$$

Test it Out

Apply the calculator strategy to any of the exercises in Chapters 2 or 7. Check your answers in the back of the book.

Answer Key

Chapter 1

#1 $x = -5, y = -6$
#2 $x = 4, y = -6$
#3 $x = -3, y = -7$
#4 $x = 3, y = -3$
#5 $x = 7, y = -1$
#6 $x = 9, y = -7$
#7 $x = 4, y = 5$
#8 $x = -4, y = 4$
#9 $x = -4, y = 7$
#10 $x = 7, y = 4$
#11 $x = 5, y = 7$
#12 $x = -4, y = 3$
#13 $x = 8, y = 8$
#14 $x = -8, y = -9$
#15 $x = -8, y = 7$
#16 $x = -9, y = -8$
#17 $x = -2, y = -3$
#18 $x = -6, y = 7$
#19 $x = -6, y = 9$
#20 $x = 8, y = 5$
#21 $x = 6, y = -9$
#22 $x = 1, y = 1$
#23 $x = 4, y = 3$
#24 $x = 6, y = -2$
#25 $x = -1, y = 5$
#26 $x = 6, y = -2$
#27 $x = 7, y = 8$
#28 $x = 6, y = -3$
#29 $x = 7, y = -9$
#30 $x = 9, y = -6$

Chapter 2

#1 $x = -4, y = 7, z = 6$
#2 $x = -5, y = -6, z = -2$
#3 $x = 2, y = 8, z = -6$
#4 $x = 8, y = -1, z = -1$
#5 $x = -3, y = -6, z = -2$
#6 $x = -4, y = -6, z = 5$
#7 $x = 9, y = -5, z = -3$
#8 $x = -9, y = 8, z = 6$
#9 $x = 7, y = -7, z = 1$
#10 $x = -2, y = -1, z = 1$
#11 $x = -1, y = 8, z = 4$
#12 $x = 5, y = 8, z = 1$
#13 $x = 5, y = -3, z = -8$
#14 $x = -1, y = 2, z = -5$
#15 $x = 2, y = 3, z = -7$
#16 $x = 1, y = -4, z = -3$
#17 $x = -4, y = 1, z = -5$
#18 $x = -8, y = 5, z = 1$
#19 $x = 8, y = -3, z = -1$
#20 $x = 5, y = 9, z = -4$

Chapter 3

#1 $x=-2, y=-3$
#2 $x=5, y=-9$
#3 $x=-4, y=-7$
#4 $x=-3, y=3$
#5 $x=-3, y=-3$
#6 $x=-7, y=-4$
#7 $x=-2, y=7$
#8 $x=-7, y=-8$
#9 $x=6, y=6$
#10 $x=-5, y=-1$
#11 $x=1, y=9$
#12 $x=8, y=-8$
#13 $x=-5, y=4$
#14 $x=-8, y=-7$
#15 $x=-9, y=9$
#16 $x=1, y=5$
#17 $x=1, y=1$
#18 $x=-4, y=9$
#19 $x=9, y=-1$
#20 $x=9, y=2$
#21 $x=-6, y=9$
#22 $x=-2, y=-5$
#23 $x=4, y=-4$
#24 $x=4, y=8$
#25 $x=-7, y=1$
#26 $x=6, y=-5$
#27 $x=3, y=-6$
#28 $x=7, y=6$
#29 $x=1, y=-5$
#30 $x=-5, y=-5$

Chapter 4

#1) 3
#2) -4
#3) 24
#4) 0
#5) 32
#6) -68
#7) 12
#8) -39
#9) 12
#10) -32
#11) 14
#12) 0
#13) -8
#14) -21
#15) -20
#16) 0
#17) 83
#18) 51
#19) 39
#20) -90
#21) 6
#22) 9
#23) -16
#24) -2
#25) -58
#26) 44
#27) -14
#28) 36
#29) 27
#30) -33
#31) 56
#32) -2
#33) -66
#34) 21
#35) -48
#36) 104
#37) 6
#38) 0
#39) 18
#40) 10

Chapter 5

#1 $x = 6, y = 6$
#2 $x = -8, y = 1$
#3 $x = -4, y = -9$
#4 $x = -4, y = 9$
#5 $x = 8, y = -4$
#6 $x = 4, y = 7$
#7 $x = -3, y = 7$
#8 $x = 2, y = -5$
#9 $x = -5, y = 2$
#10 $x = -1, y = 1$
#11 $x = -5, y = 3$
#12 $x = 6, y = 4$
#13 $x = 9, y = -6$
#14 $x = -9, y = 3$
#15 $x = 7, y = 9$
#16 $x = -12, y = 0$
#17 $x = -1, y = -1$
#18 $x = -8, y = 2$
#19 $x = -8, y = 8$
#20 $x = 0, y = -5$
#21 $x = -2, y = -2$
#22 $x = 8, y = 2$
#23 $x = -2, y = -3$
#24 $x = -2, y = 5$
#25 $x = 4, y = 3$
#26 $x = -6, y = 6$
#27 $x = 9, y = 4$
#28 $x = 4, y = 3$
#29 $x = 4, y = -5$
#30 $x = 7, y = 1$

Chapter 6

#1) 448
#2) -71
#3) 60
#4) 543
#5) 260
#6) 6
#7) -17
#8) 56
#9) -528
#10) -77
#11) -58
#12) 126
#13) -387
#14) -126
#15) 125
#16) 787
#17) 125
#18) 151
#19) -30
#20) 669

Chapter 7

#1 $x = -4, y = -4, z = -9$
#2 $x = -1, y = 9, z = -7$
#3 $x = 3, y = -6, z = 7$
#4 $x = -3, y = 9, z = 6$
#5 $x = -2, y = 4, z = 6$
#6 $x = -1, y = -5, z = 3$
#7 $x = 5, y = -2, z = -2$
#8 $x = 5, y = 7, z = 7$
#9 $x = -3, y = 3, z = -4$
#10 $x = 2, y = 4, z = -6$
#11 $x = 6, y = 5, z = 8$
#12 $x = -2, y = 8, z = -9$
#13 $x = -7, y = -5, z = 8$
#14 $x = 7, y = -4, z = -4$
#15 $x = -8, y = -3, z = 6$
#16 $x = -2, y = 4, z = 7$
#17 $x = 6, y = 4, z = -7$
#18 $x = 7, y = 6, z = -1$
#19 $x = 9, y = 4, z = -4$
#20 $x = -4, y = -2, z = 2$

About the Author

Chris McMullen is a physics instructor at Northwestern State University of Louisiana and also an author of academic books. Whether in the classroom or as a writer, Dr. McMullen loves sharing knowledge and the art of motivating and engaging students.

He earned his Ph.D. in phenomenological high-energy physics (particle physics) from Oklahoma State University in 2002. Originally from California, Dr. McMullen earned his Master's degree from California State University, Northridge, where his thesis was in the field of electron spin resonance.

As a physics teacher, Dr. McMullen observed that many students lack fluency in fundamental math skills. In an effort to help students of all ages and levels master basic math skills, he published a series of math workbooks on arithmetic, fractions, and algebra called the Improve Your Math Fluency Series. Dr. McMullen has also published a variety of science books, including introductions to basic astronomy and chemistry concepts in addition to physics textbooks.

Dr. McMullen is very passionate about teaching. Many students and observers have been impressed with the transformation that occurs when he walks into the classroom, and the interactive engaged discussions that he leads during class time. Dr. McMullen is well-known for drawing monkeys and using them in his physics examples and problems, using his creativity to inspire students. A stressed out student is likely to be told to throw some bananas at monkeys, smile, and think happy physics thoughts.

Author, Chris McMullen, Ph.D.

Improve Your Math Fluency

This series of math workbooks is geared toward practicing essential math skills:

- ∞ Algebra and trigonometry
- ∞ Fractions, decimals, and percents
- ∞ Long division
- ∞ Multiplication and division
- ∞ Addition and subtraction

www.improveyourmathfluency.com

www.chrismcmullen.com

Dr. McMullen has published a variety of **science** books, including:

- ∞ Basic astronomy concepts
- ∞ Basic chemistry concepts
- ∞ Creative physics problems
- ∞ Calculus-based physics

Chris McMullen enjoys solving puzzles. His favorite puzzle is Kakuro (kind of like a cross between crossword puzzles and Sudoku). He once taught a three-week summer course on puzzles.

If you enjoy mathematical pattern puzzles, you might appreciate:

300+ Mathematical Pattern Puzzles

Number Pattern Recognition & Reasoning

- ∞ pattern recognition
- ∞ visual discrimination
- ∞ analytical skills
- ∞ logic and reasoning
- ∞ analogies
- ∞ mathematics

Chris McMullen has coauthored several word scramble books. This includes a cool idea called **VErBAl ReAcTiONS**. A VErBAl ReAcTiON expresses word scrambles so that they look like chemical reactions. Here is an example:

$$2\,C + U + 2\,S + Es \rightarrow S\,U\,C\,C\,Es\,S$$

The left side of the reaction indicates that the answer has 2 C's, 1 U, 2 S's, and 1 Es. Rearrange CCUSSEs to form SUCCEsS.

Each answer to a **VErBAl ReAcTiON** is not merely a word, it's a chemical word. A chemical word is made up not of letters, but of elements of the periodic table. In this case, SUCCEsS is made up of sulfur (S), uranium (U), carbon (C), and Einsteinium (Es). Another example of a chemical word is GeNiUS. It's made up of germanium (Ge), nickel (Ni), uranium (U), and sulfur (S). If you enjoy anagrams and like science or math, these puzzles are tailor-made for you.

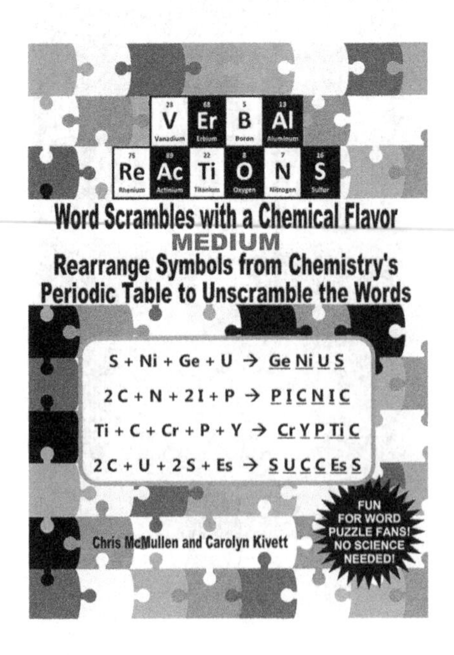

Made in the USA
Middletown, DE
24 September 2023